Warren Upton:
A Pearl Harbor Survivors Story

Warren Upton:
A Pearl Harbor Survivors Story

By Warren Upton

Edited by Don and Denise Downey

WWII Veterans History Fund
2021

Copyright © 2021 by Warren Upton

All rights reserved. This book or any portion thereof may not be reproduced or used in any manner whatsoever without the express written permission of the publisher except for the use of brief quotations in a book review or scholarly journal.

First Printing: 2021

ISBN 978-1-63010-022-3

WWII Veterans History Fund
San Ramon, CA 94583>

www.dondowney.com

Dedication

To my lovely wife, Jean.

Contents

Introduction: Jim Taylor at the USS Utah Memorial 1

Warren and Jean Interview 2015 3

USS Utah had long, distinguished history in naval fleet 29

Warren at Daughters of the American Revolution Santa Clara, CA 33

Surviving U.S.S. Utah crewmembers unite in Laughlin 35

The forgotten ship 39

Photos 43

Index 54

Introduction: Jim Taylor at the USS Utah Memorial

Start Time: 00:00
End Time: 10:57

JT (Jim Taylor)

JT: Well, that's Jim Taylor. Why walk up stairs when you can climb it. (Laughter) Any time you're in or around Hawaii and you're listening to somebody speak, the way you start your speech is I say, "Aloha" and you all say, "Aloha." Okay? Aloha.

Multiple: Aloha.

JT: That's pretty good. (Laughter) That's a lot better than I thought. (Laughter) I'm not going to have make you guys do it again because you did good the first time. (Laughter) As we said earlier, we're here to honor and recognize the folks on the USS Utah, those that were lost, those that are still inside that ship, those are buried up at the Punchbowl, and those that have been sent back to the United States, then United States and now mainland -- three of them for a proper burial that were identified. Those folks and also obviously those three that are here tonight and all the rest of them that were able to escape the ship during the attack. It's more important that we need to talk about all of you folks in here, either the family or the friends of the USS Utah. And it's not a big secret that our numbers are dwindling with the survivors and that's not fun, but it is a fact and you folks out here in this audience need to carry on the legacy that they left for you. Tom Brokaw was pretty good, "The greatest generation." The military personnel today have learned an awful a lot from the greatest generation, although they might not know it.

The greatest generation during World War II created the mold that our military personnel are using today to serve. It's great and you people that were here during the war, the three of you, and all the rest of them and your family members should be very, very proud of your either presence or absence of members of that generation that did what they did for our country. This is fantastic. I mean Pamela Calavan and her sister, Diane, this is awesome. You guys did it. You had some help, there's no doubt about it, from other people, but I can't remember ever attending a function that is as fantastic as this is. You guys really did it. (Applause)

Now, talking about the USS Utah Memorial, back about six or seven years ago President Bush signed a bill that made the USS Utah part of the National Monument, National Valor in the Pacific to be exact. Since then the National Park Service has done absolutely nothing as far as leaving it open to the public. Most of you may know that if you look up in the hospitality room, there are some copies of the newspaper article that I basically wrote for the writer. I've sent letters to the Secretary of the Interior, Ms. Sally Jewell, and Congress and Senators and everything like that and I have received zero responses, and that was about three months ago. Well, I'm not quitting. I'm going to get that darn thing opened to the public. I wish I could have had it open for Cecil because I told him I'm working hard on it. So, when I get back I've got a new letter I'm writing, return receipt request because I assure you possibly probably didn't receive my letter because I didn't get a response.

It's amazing. If you look at the picture that's in that newspaper article, the memorial had all those paint chips hanging around. It was really ugly. Well, as soon as that picture appeared in the paper, the very next day all of a sudden the paint was cleaned off and it was repainted. And the National Park Service, obviously, they said it was part of ongoing scheduled work on the USS Utah. Now, ___(04:53) any work done in eight years or seven years. Because it was in the newspaper, now the memorial is a lot better. So, I did accomplish that. I had a lot of sailors, our volunteers. They come out and clean up the area around the memorial all the time now. I even got the chaplains to go over there on their time off thanks to the ___(05:17) chapel. So, it looks good. It looks a lot better. It still needs a lot of work. And one way or the other, folks, I'm going to get that darn memorial opened to the public. (Applause) I'd like to introduce my family. I have Miley here, who has volunteered to be the official photographer of this event tonight. So, she's wandering around and taking pictures of ___(05:47) room. (Applause) And then I have my youngest daughter, Lani, and her husband, Joe, who ___(05:55) (Applause) And, of course, most of you know Nora, my wife of 15 strong years.

I think about the Pearl Harbor survivors that are with us today and we have one that jumps out of airplanes. I never could understand why jump out of a perfectly good airplane. (Laughter) But last week was ___(06:24) 94th birthday, so he decided to go out and jump out of an airplane. So, that's what our greatest generation does to pass the time. Gil Meyer, he's got six acres that he cleans up with has tractor every day. Now, anybody that wants to climb up on a tractor and get beat up by an oak tree, there's something wrong with that picture. (Laughter) He's showing everybody now. Bill Hughes, what does he do? He goes out and talks to the history classes in the local schools and that's admirable, Bill. I don't know where you are. Back there, okay. But that's great that you're doing that and if you're passing on the legacy of your generation to those students.

The Utah is getting a lot of business, if you want to call it that. As most of you know, I'm best known for returning the sailors that were here during the attack back to Pearl Harbor where they want to be after they die, so that they can be with their shipmates, their brothers. Yes, their brothers that were killed that day. I've done over 350 of them and about 325 of them were off the Utah Memorial. So, it's not that this memorial is not doing any good. I don't know if you folks know it or not, but there are a lot of graves up at the Punchbowl Cemetery that are marked unknown because they didn't know who it was. Thanks to another Pearl Harbor survivor who lives in Honolulu, his lifelong dream is to have those graves disinterred and identified with modern day technology that they could actually identify people with bones and so forth. Just a couple of weeks ago DOD says every one of those Oklahoma graves is going to be disinterred and identified. That's because of the work of Ray Emory, who is a Pearl Harbor survivor who was on the Honolulu. After the Oklahoma ones, it's going to be the Utah. I've been promised that. So, those 14 graves out there at the Punchbowl that are marked unknown are going to be known hopefully, identified. I'm going to try, and I don't know if I can be successful because it will be up to the families, I'm going to try to get those 14 put inside the ship.

Now, this is a bittersweet night for me. I think the greatest thing that -- I'm looking forward to the next Utah reunion today day after I get home from the reunion. So, for 363 days or so I'm looking forward to the Utah reunion. It's rough because I have so many people that are friends that I probably won't see again. I want to thank all of you for being here and honoring your heroes. (Applause) Most of the Utah and Pearl Harbor survivors would tell you they're not heroes. Well, I differ with that. Gil, Bill, Wally (10:21 sp?), and the rest of them that are still around and those that are gone are heroes. Anybody and everybody, including this Master Chief, and that Chief back there, and these sailors in my eyes they are heroes for defending our country and God bless the USA and God bless the United States Navy ___(10:40). Thank you.

(Applause)

(end of audio)

Retired Navy Master Chief Jim Taylor (1939-2019)

Warren and Jean Interview 2015

Start Time: 00:00:00
End Time: 01:49:00

QF (Female Interviewer)
QM (Male Interviewer)

WU (Warren Upton)
JU (Jean Upton)

WU: I was born in El Dorado, California up in the Sierra foothills. My dad was an old teaching principal and the wife actually was at that time a housewife and I was the third of four children. My dad went from one school to another sometimes just for the variety of it. And I grew up eventually in East San Francisco Bay area in what is now Fremont. My dad had passed away, so we were alone with our mother.

QF: How old were you when he passed?

WU: About seven. And I went through grade school there, elementary school in what was Warm Springs then, the little town near the border of Santa Clara and Alameda County border. And it was about I graduated from elementary school, incidentally where my dad had taught. I went to Washington High School after that and I always wanted to join the Navy. That's why I eventually did.

QF: Why the Navy?

WU: I liked the sea or maybe because of San Francisco Bay. But growing up I Fremont, I did like to go take the dog and walk down to the Bay. In fact, it was a slough, a tributary of San Francisco Bay; go swimming in that salt water. It was fun, but when I got to high school things were a little more serious. So, for my junior year the Navy recruiters came once and showed us pictures of the whole battle fleet. But in high school I was in the radio club also. I was interested in electronics and radio and about the time I graduated I was all set, but my mother had cancer and she became ill. I was going in immediately out of high school, but she said, "Well, I want you to go in the Navy, but wait a while and see how things turn out." So, I did that.

Then after she had passed away, it was the December of '37. I had already graduated high school in the Class of '37. I came across the Bay. My uncle and his wife, she was a retired school teacher, and I painted a house and she heard of a school in navigation and seamanship, so I went to that for about a year or so. And then I applied for a job with merchant marine. I was a cadet officer in those days. I got tired of waiting for that and I went up and in the winter just before Christmas to do some Christmas shopping for my sisters and brother. And I thought I'm going to go up and see the Navy recruiter.

I went in there and this chief tour captain, they don't have that rating anymore, but he said, "I'll give you one of these classification tests." And I took it and I did quite well on it I guess. So, I came out and he says, "Oh, come on. Sit down and I'll take your presents and put it in my desk drawer to keep them safe." The next thing I knew, it was as doctor giving me a physical. In those days you had to wait. They had quotas. They had to check on you pretty thoroughly. In fact, I thought afterward you'd think I was going into the Naval Academy. They checked on every school and I got tired waiting. I called. I was living in Redwood City then. And I called the chief. I said, "When are you going to call me?" He says, "Be patient, son. We'll get to you."

So, on February of 1940 I got the notice, about two weeks before that actually, to come up and I was on my way. They shipped us to San Diego for recruit training, boot camp. And I went through boot camp. It was a rude awakening like it is to all you men when you hit the boot camp, but it was good training. Looking back, it was ideal. And I took some classification tests and qualified for service school. And the service school I was approved for was right there at San Diego. We had to wait for leave after boot training. We had to wait boot leave until after you finished your service school.

QF: Can I ask you a quick question?

WU: Sure.

QF: Could you talk more about what training was like?

WU: Boot training?

QF: Yeah.

WU: Oh, it was up in the morning very early, line up for chow, and then the daily routine. They initially drilled you at infantry training marching and all the commands. They had a very definite program of training and later you had to qualify up at the Marine rifle range.

QF: What guns were you trained on?

WU: The .30-06 rifle. Actually in training station there they did what they call a small boy, the .22 caliber and we had to qualify in that too, but the .30-06. And then you got to fire two of the old Browning machine gun and the Lewis. Many of you probably don't remember those. But after I finished that I came up and enjoyed ten days of boot leave, delayed boot leave with members of my family and friends in the San Francisco Bay area. Then I reported back to San Diego and they transferred me over to what they call T-Unit, transfer unit where you were awaiting transfer to your duty station. You watched all the assignments. We read those assignments very religiously every day to see where we were going.
Well, mine came up with the fellow I had enlisted with for the *USS Utah*. And within a week or so they put us aboard a Navy oiler, the *USS Brazos* for transportation to Pearl Harbor.

QF: It was what kind of a ship?

WU: Navy oiler, a fuel ship. And so of all things I was given duty of mess cooking. That is preparation of food and everything. Fortunately, I didn't get seasick then, nor did I ever get seasick fortunately because I witnessed a good many of other people that did. But after about I think it was nine days to Pearl Harbor because they were towing a target out to Pearl that was used for practice purposes. And when I got out there I reported aboard the *Utah*. The executive officer asked me. He says, "I see you're a radio-qualified graduate of the radio specialty and I know you know semaphore and blinker-light. You know the code." So he says, "There are no openings in the radio gang. I'll give you your choice of the deck gang or you can go in the signal gang." And I said, "I'd rather go on deck, sir."

QF: Why was that?

WU: I didn't want to become a signalman. I knew if I got stuck in there --

QF: Your career path?

WU: Right.

QF: Well, how did you feel about the *USS Utah* at first?

WU: Well, it was a surprise as we thought we were going to a battleship, a real first-line battleship. But in 1932 she was decommissioned as a battleship and had been assigned as an auxiliary ship. AG was the designation, AG16. And actually it was so much training on there, the ship acted as an aircraft gunnery school to other people of the fleet, gunneries, the gun crews. And we also were a target ship. They put heavy timber on the decks so not to damage the teakwood decks so much. Then they had repair parties that would go out and repair everything between what they call bobbing waters, and you better be below out of the range of these.

At night they had smaller bombs that when they hit they flared and you could tell the color. They could tell the squadron that made the hits, aircraft squadron. Well, they had the Marine pilots, the Navy pilots, and in some instances the Army got in on it. And we even operated once as a target to submarines for torpedoes.

QF: How did that work out?

WU: They set the depth of the torpedo below the hull and they had watchmen on the ship that watched for the wake of the torpedoes and they could tell whether it was a hit. I forgot to mention how much fun it was seeing Hawaii. We thought that we had read about it so much. Anyway, after quite a number of weeks we'd come in and go out for a week, sometimes two weeks for training with the air squadrons. We came in on that Friday night, I believe it was. And all the ships, all the destroyers -- not all the destroyers, but all the battlewagons were in and a good part of the fleet. And this was more for pre-Christmas. They had contests of various bands from the different ships, boxing, and anything. The officers had their balls, as they call them.

QF: Did you see a lot of Hawaii itself?

WU: Yes, I made a point to see Hawaii.

QF: What was it like before the war?

WU: Oh, before the war it was quite nice. You relaxed. It was warm, as many of you who have been to Hawaii know. But it was enjoyable. You could swim and have a lot of fun.

QF: Do you have a favorite place you like to go to?

WU: I used to like to go swimming.

QF: You wanted the beach, huh?

WU: Yes. And also, there was a theater, Waikiki Theater. And it was the old type theater where they had a live organist, piano player before the show. And they had good shows there, first-rate shows. And it was surprising to see all these plants inside the theater. And this friend of mine, Ray Hanson, he's gone now. But he and I used to go bicycle riding out there toward Diamond Head.

QF: Could you say the name of the theater?

WU: Waikiki Theater. Yes, it was rather unique.

QF: Is it still there, do you know?

WU: I don't know.

QF: You went swimming and biking and to the theater?

WU: Yes, that's correct. The YMCA had a nice pool. You could go over there and do laps if you were tired of the beach. Anyway, it was a lot of fun before the war. And I thought I was intending to do the morning of the 7th, I got up and I was going to shave. I was reaching over someone's locker. Our stinking apartment was right next to the radio shack and you had these folding army cots. That was our bunks. You had to fold up your cot after you got up. I was just reaching over to get my shaving gear out of my locker and then this sudden jolt happened. The radioman on the watch said collision. He says on the second one, it wasn't just a few seconds later the second hit and we knew something was other than a collision because the ship began to list.

We tried to awaken everybody, but we knew we had to get out of that sleeping compartment. Where the ladders were they called air trunks for ventilation purposes and we worked our way up the ladders. And I remember that definitely that one or two fellows -- I guess that was their battle station. So, they lowered what they call battle gratings that were supposed to protect the engine room personnel from anything dropping down there. By sheer force we just pushed that up, that battle grating. It was heavy. And we went up two levels onto what was called the super structure. Ty call it the air castle. I don't know where it got that name, but it was right across the exit from the ladder and below, right across from what the -- they call it Geedanus (sp?) stand. Of course, that wasn't open.

No one was interested in that. We were interested in getting off. By the time we got up the ship was taking a list. Well, I remember there were some life preservers under and I didn't get one myself. I could swim anyway, but some of the men ran out from under the cover of the super structure and soon ran back because there was heavy strafing on the decks. I guess they were trying to keep us aboard. But I went up another deck, up another ladder close to where the captain's cabin was on that level. I went over to the side there and as I mentioned before that the uniform of the day was white shorts, like a white trouser shortened and you skivvy shirt. For undress they call it for aboard ship. I hadn't shaved yet, but I went up and over the bottom. The lines were snapping at that time as she was rolling over. And I can I remember one fellow jumped. It was a long way to the water at that time, it looked like it anyway.

He was screaming. He had a lifejacket on and then scraped on the bottom. That saved him from the barnacles because I got a few scratches, on the barnacles on the bottom of that old ship. I slid down to the bilge keel and then I jumped in the water from there and swam over -- as I started swimming, I was just swimming a sidestroke to be less of a target. And the lieutenant, he was commissary officer I remember. He says, "Can you swim, Red?" I said, "Yes, sir." He says, "Could you help me?" I said, "Yeah." So, he was in his pajamas with a lifejacket on. He was an academy graduate too, but I helped him over to the piling and the key, as they call it, where those lines went. And he tried to get me an advancement in writing later I heard.

QF: Were there a lot of planes in the air?

WU: Oh, yes. They were strafing us during our evacuation from the ship. They were strafing us heavily.

QF: It was pretty loud obviously?

WU: Yes. One of our motor launchers took us the rest of the way to Ford Island. And there was a trench that they were building to lay pipes of some type. So, we jumped in that trench to take cover because the battle was going. We saw many interesting things. I don't know the exact time that we did remain in that trench, but we saw one of the *USS Monaghan* was going out to sea and evidently they had a contact, one of the Japanese midget subs. And they made a hard right turn and they rammed it. I heard that sub was still there until they finally raised it. The thing in our favor, the guys would yell approval.

QF: How low did the planes fly?

WU: Oh, several hundred feet.

QF: You could almost see the pilots, huh?

WU: Yes.

QF: You saw the *Arizona* blow up?

WU: I heard the *Arizona*, but that was a little later. It was pretty close by the time we left that -- I'm getting ahead of myself here. The flatbed truck came and took us over toward the center of the island. And I can remember after disembarking from that truck while walking and our clothing had become saturated in some places with oil, that fuel oil. And a couple of yard workmen says, "Here, get a rag and clean." That didn't help too much and it's rough on the skin, that solvent. Then I was going a little farther and a couple of the commissary people said, "Here' some clothes. Get out of those greasy clothes, oily clothes."
As I mentioned in one of my interviews, it would seem to be they were a little large, but I was happy to have clean clothes.

Then we went over to the commissary. They were breaking out candy or anything. You didn't pay for it. They'd just hand it -- and soda drinks. I didn't get much of that. I was too busy. But one of the people from the *Utah* said that they were giving away clothes over at the ship service over here. So, I went over there and I got a pair of dungarees that fit me and a dungaree shirt so I was set, and tennis shoes as they called them in those days, sneakers. I felt a little better, but our executive officer had been ashore rather. And he was in civilian clothes I remember and he was a good officer. He was interested in taking muster and finding out who was still around. He took up to the barracks. There were the barracks over there by the second deck I remember, second story, and mustered us. We were mustered about three times after that. Then they took some of us over to the *USS Argonne*. I was a Base Force ship, a commander base ship, Base Force.

QF: Was it in dry docks?

WU: No. It was in dock. It was more or less with a command aboard it. It didn't sail too much.

QF: Where was it located?

WU: I know on another part of the island, Tin-Tin Dock (31:29 sp?) I believe they called it. They issued us gas masks. First I thought I'd take a shower and get the oil and stuff off, and I used my underwear as a towel because I didn't have it and put my dungarees back on. While I was taking a shower a fellow from the radio gang, he says, "What is your rating?" I said, "Radioman Third." He says, "We can use you up in the radio shack." So, I remember the chief radioman was the name of Galli (sp?) was his last name. They put me on watch there that first night. These cuts and bruises on my legs from sliding down the barnacles were beginning to get pretty sore. He could see I was squirming a little.

QF: Were the cuts just on your legs?

WU: Yes. Yes and a little higher. (Laughs)

QM: That's what we were asking. (Laughs)

WU: I neglected to say we had handled a little ammunition unloading that night before the -- we went to turn in. But the Chief Galli said, "You better go down and turn in. I know you're pretty tired and your legs and bottom seem to be a bit sore." So, I did. I went down and just as I got to sleep I heard firing of machine guns. We jumped up and ran up to the mess deck and there was a big coffee urn up there, which was a large one. Anybody could draw a cup of coffee. And we were sitting around after that just momentarily and the machine gun fires began. And a .50 armor-piercing cartridge came through the side of the five-inch hull of the *Argonne* and it hit and killed a yeoman striker from the *Utah* after he had gotten off of there. It was a double tragedy, friendly fire. They were firing at something. In fact, it was the *Enterprise* planes that were coming in early and they fired on those. Three of them crash landed. I don't remember if any of the pilots were killed. At that time Ford Island was a naval air station there. Of course, they were all prop planes. No jets at that time. After this excitement was over I went back to sleep.

QF: Do you remember the name of the man who was killed?

WU: Not offhand. I have it somewhere.

QF: Were you there when it happened?

WU: Yes. Yes, not too far from the poor guy.

QF: It could have easily been you?

WU: Yes. You wonder if your number isn't up. It isn't up, I guess. But he was from the first lieutenant's office, a yeoman. Anyway, the next day, the 8th, which was on a Monday, we sat around, most of the radiomen, and talked about what happened. I guess that was our way of getting rid of the anxiety. I was assigned to mid-watch on the bridge, radio telephone. The *USS California*, of course they were abandoned and there were probably more on Ford Island than on the ship. And the Commandant Navy District 14, most of my mid-watch was taken up by testing, routine testing. There wasn't anything that came up. The only thing I got about the sunrise was that the planes would be taking off from Ford Island, the *USS Enterprise*. And the ship had come in in the meantime and they were following destroyers that were sweeping the channel for any midget subs or large subs.

After the mid-watch I turned in; had breakfast, turned in. Just waking up, not quite awake and a seaman came down from the communication office on the *Argonne* and he says, "You and Gross." He was a striker from Berkeley, a radio striker apprentice and one of the communications officers. I went up there and our Chief Putman (sp?), our chief radioman from the *Utah*, he knew I lived in the Bay Area, so he arranged for me and Gross to get on the *USS Castor* for transportation back to the States. They were going back. In fact, we had taken aboard four of the last Pan-Am radio operators from Wake Island that had gotten off there as passengers. The guys, they want to stand on the international distress frequency. We let them. They were real nice guys. We often wonder what happened to them.

We arrived back in the States. It was about five days later I guess, pretty close to Christmas Eve. After the attack, I forgot to mention that, they handed us postcards, multiple choice, "I'm wounded, I am all right" to send back. Those cards got back 30 days later, so I beat the card back.

QM: Do you remember going through the Golden Gate or were you on deck when you came back to San Francisco?

WU: I was on watch. I was in the radio shack.

QM: Oh, so tell us all about that and how you felt coming back home after.

WU: That was great. But unfortunately, there were other ships with wounded that arrived about the same time. I sent word over to one of my sisters that I was aboard and she worked for the phone company at that time. What I did though, I took Christmas duty on Christmas Eve and Christmas Day in exchange for a 72-hours after that. I remember meeting my sister. I was so happy to see her. From then on I was on the *Castor* for a while. They went in the Navy yard in Mare Island and I was second class then. They transferred the first class off and the chief, so I got stuck with Senior Petty Officer for the radio gang. But from then on during the war it was a lot of happenings, ups and downs.

QF: Well, go ahead and talk about them. So, you were on the *Castor*?

WU: Yeah, I was on the *Castor*. Well, went to the Navy yard and got some new communication gear and they upgraded some of the gunnery. We were there for about a week it was and then we sailed out to Pearl again, Pearl Harbor. And we got out there and we had sealed orders. Actually, we were going down -- the name slips me now. I'll remember it in a minute. It was down near Tonga. We sailed with the ___(52:00) convoy with the Crescent City 4th Marine Defense Battalion was onboard that, and the destroyers *Henley* and *Helm*. I think that was pretty close to all the ships. We arrived down there and the Marines were going down there to defend that island.
After that we sailed to so many islands, Makin Island, Bora Bora, you name it.

QM: Well, why don't you take your time if you're willing to and just tell us anything about the islands, what a typical day was like, what you guys were doing. Did you shoot your guns off? Tell us about the radio work. This is really the interesting part of your Navy career, so thank you.

WU: Well, every once in a while when we were underway we would have a submarine scare. The ship could do about 30 knots. In fact, we sailed first out of Pearl Harbor for the States, we sailed without an escort. There wasn't any escort at all.

QM: So, keep going. This is great.

WU: Could I take a break a minute?

QM: Well, we can. And when we come back, we want to hear about did you ever cross the equator?

WU: Many, many times.

QM: You know what we're talking about?

WU: The International Date Line.

QM: The Golden Dragon ___(54:20). So, we'll pick it from there in just a minute.

WU: Okay.

QM: This is getting interesting.

QF: Okay. We're back for segment number two. And we're talking about you touring the island in the Pacific.

WU: Yes. Of all the islands I visited, Tarawa was the worst. There was on palm tree practically standing after that battle. It was quite an awful battle, one of the heaviest bombardments during the war except for ___(56:31).

QM: Were you there for it?

WU: No.

QM: You just saw the aftermath?

WU: Right afterwards, yes. We visited Marshall Islands, Espiritu, New Hebrides, and Eniwetok. Of course, that's part of the Marshall -- the Gilbert Islands.

QM: Tell us anything about all of those islands. Wow, it's way out there. That's one thing I know. But take a minute and talk about that. Thank you.

WU: One of the interesting ones that the destroyer tender USS Dixie was stationed near New Hebrides. It set up a nice swimming area. This was after bombardment of course, long after. And had a nice picnic I call it. Nothing compared to the luxury of the present day picnic areas. And of course they had a band I guess while they were stationed there. But we went swimming of all times on Christmas Day. I can remember that so well. What a relief from the daily routine. Of course, the daily routine of a radioman was you stand watch, as they call it, up to six-hour watches rather than a conventional four that the deck force stood. Sometimes in a hostile area you would have to stay on watch. And if you went to general quarters you had to remain at that for a few hours.

It was one of the most rewarding jobs I think. We needed communications. The timing was a bit hectic, but as long as you did your job you were -- you got to deal with dispatches quite a bit; take something up to the different officers. But those days were so rewarding looking back. You weren't sure what was happening once you turned in at night whether you'd wake up for a watch or wake up in the water, but that's part of life. I had enlisted for six years. I was over in Okinawa when the war ended and got back a bit until my enlistment was up in February. I got back in March and I wasn't sure I was going to stay up --

I had looked into a job or two. I got one at Mackay Radio in San Francisco. I had joined organized reserves, meet on Treasure Island and go on two-week cruises. I was down in Santa Cruz with a friend, a young lady, and went up to get something to eat. I heard the radios and they said that South Korea had been invaded. I said, "Oh, here we again." And by September I was back in for Korea. So, I spent five years in the reserve and about a year-and-a-half over in the Korea area. It was altogether a different type of a war then. It was only five years later, but I wouldn't trade my Navy experience for anything. It's something you grew up with and your shipmates you never forget. Most of mine are gone now. In fact, the fellow I enlisted with, it was through him that I met my wife, Jean. We still keep in touch with their daughter. In fact, she'll be coming over next week for a visit. As they say, there's nothing like comradery of the service. What we did was what anything would do with no great heroism or anything. We just did what we had to do. I would never trade the experience. One of the best decisions I made to join the Navy.

QM: Well, let's do take a minute here and talk about -- you just talked about your buddies. Just for a second, any of the buddies that you want to talk about and maybe expand a little bit about their character, just anybody you remember from the service. We'd like to hear. And did you ever get any trouble?

WU: Well, on the *Utah* when I was on the deck force one of the Boatswain's Mates Second Class comes up to me and he says, "Put a little weight on that scrub." We used to scrub the decks naturally. "You are not paying your way." I said, "Hell, I paid that long ago." And the next thing I knew I was down on the double bottoms chipping the paint.

QF: For being sassy, huh?

WU: I found out you don't talk back to a petty officer.

QM: Okay. That's a good one. We want to hear more about what it's like to be a radioman. I know for you it's just part of the business, but I have no idea about the equipment, the frequency, the call signs. Can you explain that to a layman?

WU: Sure.

QM: Thank you.

WU: Sure. Being a radioman was complex. On the *Utah* before the war we stood communication watches, they call it, and then you had to do the technical work on the odd day. You had to repair radios. There was no electronic technician at that time. They came in after the war. And in fact, some of the radiomen went to that.

QM: A lot of tubes.

WU: Vacuum tubes, sure.

QM: Well, I have some of those vacuum tubes in my stereo at home. But yeah, keep telling us more details about the equipment and then get into what you did. Thank you.

QF: Specifically on watches, yeah.

WU: On regular radio watch you had a typewriter and a position and you typed that as you listened. On the International Morse Code they sent it. You had headphones on. They used to call that the mail. I remember the typewriters. They were all Underwood's and they were good. I've seen them fall on deck and crack and then weld and frame. Of course, there weren't a lot around during the war. In fact, the radio gangs used to exchange Christmas cards before the war; that comradery amongst the different radio gangs. I remember a lot of the call signs. The call sign for the *Utah* was NIQJ, and the *California* was NAFT, and *West Virginia* was NEDJ. I don't know why those stick with me, but they do. I'll never forget them I guess.

QF: So, did you send out messages?

WU: Yes. They had what they called -- for general information they sent them automatically with automatic senders, they call it, but transmit. And that's what we copied, the fox schedules they call them. And then you had to communicate manually with a telegraph key. If you became real proficient, I still have my old speed key. They called it the bug in those days. But it was so much fun in a way. And you better not miss too much while you were on a circuit. That first radio telephone was on the bridge mainly. And on the *Utah* we had the aircraft set on aircraft frequency that we communicated because I didn't cover that part where they would have a drone similar to today. In fact, they could put the Utah under radio control, complete radio control. That would operate the engine room and steer the ship. There was one Radioman First Class was a specialist. He's gone now too. He moved back to Maryland. But it was kind forego of the present-day drones.

QM: That's so interesting. If there is some way you could just describe the equipment or how that worked or anything about that system, we'd like to hear it.

WU: I don't know too much about that particular system. I know it was naturally a combination of vacuum tubes and receivers. Of course, the transmitter from the destroyer would control the ship with their transmitter and a set of servos, and relays. It was a crude type for that, but it was something. They controlled the World War I plane they rigged up once during. It would run radio control and that was top secret more or less. It crashed. They shot it down. They crashed at sea. They retrieved it and posted kind of a security guy so no one would get around it. Those were the days. As they say, they're a forerunner to our present-day electronics to the solid state. I later worked as a radio tech for United Airlines, so I got to see from the vacuum tube to the solid state.

QF: Could we go back to the Navy just a second?

WU: Sure.

QF: So, when you crossed the International Date Line and got your Golden Dragon, did they do anything?

WU: Not for that, but I remember the initiation for crossing the equator.

QF: What did they do to you?

WU: Well, they'd paddle you and whatnot and put grease on you.

QF: Which ship were you on at the time?

WU: I was on the *Castor*. But because of the security and the idea that you could be torpedoed, they limited that. Nothing like you hear some of them went through, but it was enough to know that you were crossing the equator.

QM: Yeah, exactly. I think the ceremonies were a lot more elaborate before the war.

WU: Yes.

QM: Where they were down there looking for Amelia Earhart or something. But go ahead, Denise. We've got a few more things maybe to wrap up the war part. And then we really want to hear everything after the war. So, maybe we can come back to hear a little bit more about the Navy. But, Denise, why don't you go ahead and put a wrap on the Navy?

QF: So, after the war you were out for a while. And then the Korean started and they pulled you back in?

WU: Yes.

QF: Okay. What was that like?

WU: It was entirely different. In that five years I think they had let down the guard on training to a certain extent. There were a lot of new personnel in the service. Some had stayed in. But the training wasn't there that they had before the war was over. You knew what to be careful about. For instance, the ship I was on was firing on North Korea, firing on the bridge up there. I was the radioman in charge then. I was Chief Petty Officer and I walked in and I said, "Close those ports and get down below because they might fire back." That was common sense. Everybody knew that during the war, but it was unusual. I'd seen the carriers from their conversion from when I first went over there was from the old piston type plane to the jets. Those poor jet pilots, more than one of them had to ditch, but that's where the choppers came in handy

16

pulling them out. But around October it would start to get cold off coast of Korea and never lets up until the spring.

QF: But which ship were you on?

WU: Let's see. *Burrows* I believe.

QF: *USS Burrows*?

WU: Yeah. It was a different war, let's put it that way.

QM: How was the equipment? Was the equipment worse off or did they have weather equipment for you I hope.

WU: Well, the communication equipment was better naturally. There was more radiotelegraph, as they call it. Now the radio rate is no more; it's information technician. But I'm still in touch with people out there at Pearl.

QM: Anything else about the Korean War? Of course, it's still not technically over.

WU: Yeah. As I was going to mention, they had pushed the snow and things off with a little caterpillar before the planes would take off. That was, as I say, some war. I saw the Incheon, Wanson. We went clear up to where the *St. Paul* was operating. Speaking about personnel, there was a fellow. He was an electronic technician during World War II. He had gotten out and he gone to UC Berkeley and got his commission. He was an Ensign. He had a family in Berkeley and he came out after I did and I showed him around the ship. We were going to transfer him at sea to the St. Paul. And the poor guy was doing some work delivering. He was in charge of well boat delivering some relief stuff for some Korean natives and they hit a mine in the harbor. I found out when I got back that he was killed.

QM: Yeah. Those are the stories you remember. Wow, pretty sad.

WU: But that's part of the service.

QM: Maybe thinking of the big picture about kind of US Foreign Policy and the generals and the admirals, what did you think about people like MacArthur or Halsey, Eisenhower even? Anything you want to say about just the politics of the time or they how conducted the different wars?

WU: Well, they used to joke about MacArthur actually, but he knew the Orient, that's for sure, the Philippines. He did a great job, but you hear different things. I'm not going to put any of them down. They tried to do their job. Nimitz was a great admiral. That's one thing.

QM: Yeah. Expand on that real quickly. What do you say that, Warren, since you were there?

WU: Well, he did a great job of doing things. I understand he was the one that started lighting the beacon over there.

QM: Yes.

WU: And Jean was a Navy nurse. She said that his wife was instrumental in doing things for the personnel.

QF: Yes. And again thinking maybe more about the big picture, you were young enough of course to remember President Roosevelt. How did you feel about him, Truman?

WU: I thought Roosevelt did a good job. I think when he passed away there was quite a general nonbelief, you might say.

QM: Yes, I've heard similar things. He was almost like a father figure.

WU: Yes. Well, with all these programs, some of them today over there in San Francisco and Hoover Dam and those places.

QM: And something we'd like to ask about, but it almost kind of sad to me. It's going to be the anniversary of course coming up of the end of the war. But especially someone, you being out there on a ship in the middle of the Pacific, how did you feel about the use of the atomic weapons? Thank you, Warren. It's a tough question.

WU: It was regretful that it more or less had to be used to end the war because the Japanese were not about to give up. They were not. And there would have been heavy casualties on both sides. I know I was in Japan during the Korean War. In fact, I saw at a distance from the train that transported us down to Kyushu from Yokosuka. And the Japanese conductor says -- you could see, as I say a distance, but it just looked level, real leveled. As I say, it's regretful and I just hope they never have to use that again. I can see why they don't want nuclear war capability with every nation.

QM: Especially in the Middle East I guess.

WU: Yes.

QM: You bet, Warren. So, let's keep talking and wrap up your Navy career and kind of why you left the Navy. But then we're going to sit back and if you can muster the strength, we want you to just talk about your life after, your jobs. That's even going to be more interesting, so maybe I'm just going to sit down and shut up while you just continue on. That's going to be super interesting coming in, like you say, from vacuum tubes to solid state, all this radio work. So, thank you.

WU: I wasn't sure. In fact, when I went out to Shoemaker, California to get paid off -- I forgot to mention how I became chief. I had completed the courses and this old chief warrant, he says, "Red." I had red hair at one time, believe it or not. He says, "Why don't you go up for chief? You have enough time in in first class." I said, "Oh, I don't know." He says, "Well, you get paid for uniforms and everything." The next thing I knew, I was chief.

QM: Wait. This is really a big story because everyone -- this is a big thing to become a chief unless you've been to college and you're an officer or something. So, back up and let's --

WU: Well, there's a lot of schooling the Navy, believe it or not. There was at that time. I filed something, several applications. And as I say, I had the time in and everything and the background and I was happy I did; a few more privileges as chief. You have your separate mess. They have the chief's club ashore. In fact, they used to have pinball machines, you might call them. (Laughs) It's good to go up the ladder. But I would never trade my Navy experience for anything. It's something, as I mentioned before, the comradery amongst shipmates and even other naval personnel you might run into. But there's a lot of difference between peacetime and wartime.

QF: So, what was the year you got out of the Navy?

WU: The first time was March of 1946.

QM: Korean War you left again?

WU: Yes. Harry Truman extended all the enlistments for a year or so. I put in my year for Harry, so I got five years of reserve time. I was almost went into the Naval Air over in Alameda. And I went in there. I was going down to visit my sister that was living in Redwood City with her husband at the time. I stopped in there on the way. It used to be the Oakland Airport. The chief over there says, "Hey." He says, "You come on." In fact, I had received a letter from him after. He says, "Come by." I got busy and never went back unfortunately or fortunately.
I went and worked for Mackay Radio, as I mentioned before. And that was some operating; radio teletype, you name it, but it was good.

QF: Did you work for them after the Korean War too?

WU: No.

QM: Where were their offices based?

WU: It was 22 Battery Street in San Francisco. And I lived there in San Francisco, so it was pretty close.

QF: So, what did you do after the Korean War?

WU: Well, I had went to work for this old shipmate of mine that said the Bank of America was hiring because they gone out on a strike at Mackay at that time. So, I went over and got mixed up in banking. I worked for a couple of -- went to some night school after they had courses, American Institute of Banking courses. They were quite informative. They had old Dr. Cross from UC Berkeley teaching economics.

QF: Did you get a degree?

WU: Never did, not quite.

QF: How long did you work in banking?

WU: About, let's see, close to ten years I think.

QF: So, now you're up in the 60's to 70's?

WU: Yeah.

QF: Okay. Why did you leave banking?

WU: To go back into electronics were more lucrative you might say, a little more money.

QM: So, you're banking but then you're thinking electronics is obviously more lucrative.

WU: I believe it was more my --

QM: I think you liked, yes. Keep going. Tell us all about the switch.

WU: Well, I was on military leave, you might say, for the Korean War. And I went back there and I went to work for Crocker first and then I worked my way into the investment department. That was one of the first banks, they all do it now, but involve in investments. We worked with the brokers along Montgomery Street. I thought Navy parties were drinking.

QM: Well, really quickly. I'm imagining you're wearing a suit and tie and a hat?

WU: Right. Not a hot.

QM: You didn't wear a hat?

WU: No, just a suit and tie.

QM: Okay. Keep going. Thank you.

WU: But a lot of high stress in financial, but it taught me a bit more about accounting, which has always come in handy. Nothing is lost. I worked for Eitel-McCullough for a while. I worked for Beckman Instruments. I worked for the City of Palo Alto in their electronics maintenance. Also, I retired from the County of San Mateo as similar. I got to respect what the poor law enforcement does when working with them. One of them, he was with a K9 unit and that poor dog would get up. He'd hear a siren and he'd howl. One of them told me the dog takes on the personality of his trainer, you might say. But I enjoyed working at that. I decided to retire when I was about 67. I worked part-time after that.

QM: Yeah. Tell us a little bit more about your interactions with the police force. It sounds like there might be another story there somewhere. Can you think of any? Take a minute and think.

WU: No. As I said, I got to know some of their -- every once in a while you get to meet their family. They're like anyone else; they have their job to do. And going through the police academy is no picnic. Our oldest daughter had to do that for her state job.

QM: Well, I probably shouldn't ask this. So, you got good benefits from the City?

WU: Yes. Yes. Yes.

QM: Well, good.

WU: Very good.

QM: So, again a government job.

WU: Yeah. But if you're conscientious and do your job, you really can feel rewarded.

QF: So, I'm going to go back. So, what year did you get married?

WU: 1958.

QF: And where did you get married?

WU: San Mateo, California.

QF: And how many kids did you have?

WU: Five.

QF: How many grandkids?

WU: Four.

QF: Any great-grandkids yet?

WU: One.

QF: So did you have any hobbies or anything that you did after you retired?

WU: Well, I got on the internet for one thing. I don't know if you'd call that a hobby. And actually I use it for emails with old shipmates, the ones that have that capability.

QM: Tell us all about that. Is there still a *USS Utah* Association?

WU: Not really. And there's one more. Within a week or two there's going to be one in Las Vegas. I won't be able to make it naturally. It's supposedly the last one. This fellow that passed away right after the reunion, he had quite a large family and they worked with him and they were a nice family.

QF: Do you belong to any other associations, military or -- ?

WU: I belong to VFW. I haven't gone to many of their meetings. I support them with raffle ticket donations and I get their magazine. They have a lot of good information there.

QM: Now, in the past the Pearl Harbor group was bigger and a lot of these other groups were bigger. Do you remember any reunions from the past that were memorable from any of your organizations that you'd like to tell us about?

QF: One at Laughlin?

WU: The Laughlin? That was nice. There's a fellow. He's a Navy veteran of later years that had a dedicated, near his home, a memorial to the *Utah*. Caldi Nicker (sp?) is his name. He has a son in the Navy now, but I keep in touch with him. And his children were in the scouts and they were at the presentation of the colors and whatnot. There are so many. When our guys were, when my shipmates were able to, we used to put on reunions at the home area of the survivor. So, we've had two in Texas. We had five in Salt Lake City. They treated us so well there. That was the first one. It was in 1988 and we liked it so well. In fact, I was secretary for years. I used to make all the arrangements and correspondence with the hotel.

QM: I'm going to give you these and maybe you can just show us a few of these pictures real quick. No big deal, but just there are some reunion photos with you. Is that a photo of you in black and white there?

WU: Yes. This is me when I was Radioman Third Class. You can't see the hat very well there, but I had a white hat on. And that's the one I have hanging up in our bedroom.

QM: Well, you know? You look incredibly happy. Your childhood had to be a little rough. I know you didn't really get into the emotional side of it, but you lost your dad and your mom. That was pretty shocking to be listening to it here. You look like a happy sailor there. Like you had found your place, is that right?

WU: That's right.

QM: You found a home away from home, wow. Sure, go ahead.

WU: This was taken of some of us at Salt Lake City. There were about five of us I believe that made that mini reunion, you might say. We had five regular reunions in Salt Lake City. They treated us so well there. This one is taken of me and a fellow from our Pearl Harbor group, Ted Ivy. He's gone now too. They invited us out there after December 7th.

QM: And this one is one of your photos just showing the current condition of the memorial there. Why don't you talk about the memorial? You're a survivor. I know you're

	probably very happy that there is a memorial, but personally I could see it maybe in a little better condition, a little bit more well-visited. But what do you think about it, Warren?
WU:	I think it is sadly neglected. The gentleman who has been a champion for us out there that's in charge of burial of ashes, scattering of ashes at Pearl Harbor, he's been one of our champions. In fact, I recently asked him to escort a family on our block over there and he did it willingly. I got in touch with him by email.
QM:	Yes. Well, I'm glad you have a champion there so you have something.
WU:	Yes.
QM:	Some contact there. And thank you for sharing that, yes.
QM:	How often have you been back to Hawaii?
WU:	Not too often. We were back for the 60th. There is a brochure there from the 60th reunion. This 60th reunion, on the evening of December 6th we had a sunset ceremony there with the Commander of the Hawaiian Navy Region. And he did us the honor of (riding a white boat?). We went around. And then they sounded Echo Taps, which is always very touching.
QF:	Have you ever been back since?
WU:	No.
QM:	So, I guess why don't you go ahead and you can ask the questions about this one?
QF:	Okay. You could point it. Where on the ship were your sleeping quarters?
WU:	Right under this part here.
QM:	Actually circle that and then well draft it up for everybody. So, take your time. And Warren is going to circle that and thank you. So, I guess when you were on the *Utah* the big guns had been removed. Is that correct?
WU:	Yes.
QM:	So, did you hear other big guns going off during your time in the Navy?
WU:	Oh, yeah.
QM:	What was that sound like? I'd like to just digress and hear what was that like?
WU:	It's something that is a bit disturbing to your ears.
QM:	That bad, okay.

WU: Yeah.

QM: And would you have hearing protection?

WU: Sometimes, yeah. Actually, it was more of a boom than the crack of a five-inch is something that is really disturbing, more disturbing to the ears you might say.

QM: Okay, yes. I guess you would have told us if you had seen one of the atomic bombs or had been involved in that testing.

WU: No.

QM: Sure. Well, great. Well, go ahead and show us the *Utah*. Thank you, Warren.

QF: So, then we wanted to know where your battle station was.

WU: It's it right in the --

QF: It's the same place?

WU: Yeah, in there.

QM: Okay. When it flipped over what part of the hole did you slide down?

WU: Right about here.

QM: Barnacles were down there somewhere.

WU: Right here.

QM: Go ahead and circle that again. Thank you. We'll fix that up and draft that up. And then the other thing we're going to do is make you a really close-in map of where you were. This map is fairly detail, but we're planning on making you one, a very large like we used to do posters. It's just this area. Now, maybe we can just have you point. So, you were over at the *Utah*. And then I think you ended up over here at Tin-Tin dock?

WU: Right.

QM: Sure. Thank you, Warren.

WU: You're welcome.

QF: I was actually curious. So, you have five kids?

WU: Right.

QF: What do they do now? How many girls and how many boys?

WU: Three girls. Our youngest daughter, she teaches. When she's isn't involved with cats she teaches over at West Valley College, philosophy. Let's see, Mary works for the State, an investigator for the DMV. And Chuck works for the State. He's a software engineer for -- I'm trying to think of the department. It has to do smog control and stuff for that and efficiency; they more or less police little garages and things when they get too many complaints.

QM: Right.

WU: BAR, Bureau of Automatic Repair.

QM: Amen to that job. Okay, you're famous for this ship. And that's quite a picture there. But here is the really important person in Warren's life right here and we're going to talk to her next. So, why don't you describe that person, Warren?

WU: Oh, she's been my executive officer. (Laughs) She's great, the greatest.

QM: Well, let's go meet her. So, we'll kind of stop now and we can come back and everything. Is that okay, Warren?

WU: Sure.

QM: Okay.

WU: If it's okay with her.

QM: So, Denise is just going to say hi and this will be so much fun. And, like I said, this is the uplifting part. Who cares how many Japanese they killed? Who cares? Well, the good thing is they defeated Nazism and that was pretty good and we have freedom and democracy and religious freedom, so that's nice. But aside from that, the happy endings are so great. Hey, they went home. We went home. They have families. We have families. Think of what we accomplished since then. It's unbelievable, right?

JU: And our world is a mess now.

QF: Well, it could have been worse.

JU: Our world is a mess.

QM: I watch this stuff on the news pretty religiously myself, but I've been over there enough to know these people are just so ignorant. I mean it's not like these guys went to college and learned.

JU: What people?

QM: Just all these people shooting each other over in the Middle East.

JU: Oh, yes.

QM: I mean that's bunch of kids. You those kids running around in Africa and Nigeria, they're 12 years old, 15 years old.

JU: It's sad. It's sad. We lived in the great generation.

QF: Yeah, you were the greatest generation.

JU: That's it. You know, Tom Brokaw in his book?

QF: Exactly, yeah. So, here you are with your better half. And her name is?

WU: Jean.

QF: Jean, okay. And you've been married?

JU: Fifty-nine.

WU: Fifty-seven years.

QF: It feels like 59, huh?

WU: (Laughs)

JU: Fifty-seven.

QF: Fifty-seven years. And you have how many kids?

JU: Five.

QF: Oh, wow. So, Jean, we haven't heard anything from you. We'd like to hear about your childhood and the whole story.

JU: Oh, really?

QF: I understand you were a nurse too?

JU: Yeah, I was a nurse.

QF: Okay. Well, you could just tell us briefly.

JU: Yeah, I'll make it brief. I grew up in Oregon.

QF: Which part?

JU: Well, I was born in Forest Grove, which is just out of Portland. I went to nursing school up there. And after I was in nursing school, shortly after I joined the Navy. The war was just beginning to wind down and nearly everybody in my class joined the

Army or the Navy. The nurses that were just graduating, they weren't staying home and taking care of the people at home. They were gone, you know? I was just in the Navy two years as a Navy nurse.

QF: And where did they send you?

JU: I spent the time in California. I kept asking for hospital ship duty, but I didn't get it.

QF: Which part?

JU: I wasn't in it long enough I guess.

QF: Which part of California?

JU: Here in Oakland in the Oakland Naval Hospital and then over to Southern California to the Marine Corps base. They have dispensaries all these places. And up in San Pedro the dispensary there, that's where I was when I was in my final duty.

QF: Okay. And how did you meet that guy?

JU: How did I meet him?

QF: Yeah.

JU: That was a long time later. We got married when we were 38 years old. I mean it wasn't this early childhood kind of marriages that a lot of people our age have started out with, but I met him through a group I belonged to in our church. The lady was my friend and her husband was Warren's best friend in the Navy and they thought we should get acquainted since I had been a Navy nurse and that we had something in common. And we did have something in common. We got married.

QF: Yeah, there you go. Now you've got a lot in common.

JU: Yeah.

QF: So, any vacation you guys took together? Where all have you travelled during your married life?

JU: Most of it was to *Utah* reunions.

QF: Well, at least five. Did you go overseas at all, Europe?

JU: You mean travelling?

QF: For vacation, yeah.

JU: No.

QF: So, you've travelled around the US?

JU: Yeah, to some extent. We took some vacations, of course, with our kids; Yosemite and places like that, you know? But we haven't really done that much travelling except with the *Utah* reunions, and up to my family in Portland and around up there and that kind of thing. So, we've had a pretty simple life actually. We've been busy with five children for a long time. They're grown up of course now and the grandchildren are grown up too.

QF: You've got how many great-grandchildren?

JU: We have one little great-grandson who is adorable.

QF: How old is he now?

JU: He's a year-and-a-half. He comes over once in a while and we have a lot of fun with her. Even our caretaker adores him. She's trying to teach him Spanish.

QF: Hey, being bilingual is good.

JU: Yeah, his mother wants him to be bilingual. Our life has been pretty simple really. It hasn't been a lot of big excitement.

QM: We have about three or four minutes left, but I want to hear about your childhood before the war. Were you on a farm or the city or where did you grow up, please?

JU: I grew up in Portland, Oregon. Well, more or less.

QF: Kind of on the outskirts.

JU: Actually, we lived up in Kelso, Washington, a little lumbering town when I was little and I finished high school in Portland and went through nursing school in Portland. And then after nursing school, of course, I went into the Navy shortly after that.

QF: Do you keep in contact with any of your nursing friends?

JU: Pardon me?

QF: Do you keep in contact with any of the nursing friends?

JU: They're all up there?

QF: All of them? Okay, you're the last one left.

JU: A few. I have one friend that lives across the Bay that has nine children and she's my age. And she's not in the best health, but her kids are all there for her. She says, "That's why I have so many kids." They're all there for her. Our kids are always there for us when we need them. We can't complain at all.

QM: Is there a lot of comradery amongst nurses like there is amongst Navy men?

JU: Oh, yeah. People in the Navy were the best friends I've ever had.

QF: Do you have any quick nursing stories, any cases that really stood for you?

JU: Not really. I went on to do public health nursing.

QF: Any stories with that, any cases?

JU: Well, we helped organize well baby clinics up in the hills in Oregon and that was a lot of fun. And we had school immunization clinics and we visited mothers with new babies and that was what my nursing career was after. I went back to school after I got out of the Navy and they paid for that of course.

QM: Well, let me tell you how I feel about nurses. If there was a nurse there and doctor standing next to her, I would go with the nurse for healthcare.

QF: Do you want to get Roxie in real quick?

WU: Roxie, do you want to jump up there?

QM: Do you want to see if she will jump up there?

JU: She's not allowed to.

QF: Oh, no. She won't do it, okay.

JU: I don't know if she'll do it or not.

QM: We'll turn it real quick, okay.

QF: You can barely see her there.

JU: Roxie. Come here, honey. Come on. You can come up.

WU: Come on.

QF: She's like, "No, you're trying to trick me."

JU: She's been reprimanded for getting on the couch.

QF: Okay. And she's how old?

WU: Eight years.

QF: Really? Wow. And she's your daughter's dog?

JU: Our oldest daughter.

QM: Now, believe it or not, we actually have a special request. It's a little unusual. But, Denise, why don't you do the special request?

QF: Oh, I was wondering if you can just give each other a hug and a kiss.

JU: Oh, of course.

QF: Aww. Well, great. So, anything else you guys would like to say to wrap it up?

JU: No.

WU: Nothing like the service to grow up with in a hurry.

QF: Yeah, I bet you matured fast that day. And go Giants!

WU: Go Giants!

JU: We love the Giants.

WU: Yeah.

QF: I could tell.

JU: I don't wear this too often. I think it's too exhibitionist for him.

WU: It's a conversation piece.

JU: I wore it to the doctor one day and even the ladies were saying, "Go Giants!" I think this will be something you could give to little Haley, our little granddaughter and maybe the teacher would let her present it. Kids don't know that much about Pearl Harbor. We have a little granddaughter that's in the 6th grade and I think she might be interested in asking her teacher if this could be presented to her class.

QF: Yeah, that would be great.

JU: At least a portion of it, about the Navy and stuff.

QF: Yeah, where does she live?

JU: Hollister.

QF: Okay. So, it's not like you could just run down there and give a talk.

(end of audio)

USS Utah had long, distinguished history in Naval Fleet

By Joseph Bauman for the Deseret News

Warren Upton will never forget the ship named after this state. He managed to get off the USS Utah after the ship was torpedoed at Pearl Harbor on Dec. 7, 1941. When he finally climbed the ladders and reached the main deck that Sunday morning, he said, "The ship was beginning to list a little more" and Japanese planes were "strafing heavily."

Upton, now 89, had planned to travel from his home in San Jose, Calif., to Salt Lake City for a reception to be held on March 9 marking the 100th anniversary of the laying of the battleship's keel. However, he said, his wife has a medical appointment that will prevent his attending.

William Hughes, 87, another survivor from the Utah on that 1941 day of infamy, will be at the reception, scheduled from 5 to 6:30 p.m. in the Capitol rotunda. Hughes, who lives in the Dallas-Fort Worth area, told the Deseret News that he believes about 40 survivors may be still living.

The USS Utah Association held its first reunion in Salt Lake City in June 1988, Hughes said. "I cannot explain what a thrill it was seeing old shipmates," he said. Other reunions also were held in Utah, he added.

An exhibit of art, photographs and artifacts of the USS Utah will be available for public viewing for several months starting March 9. The display, planned by the Capitol Preservation Board and the Fort Douglas Military Museum, will be on the fourth floor of the Capitol.

The reception and exhibition are sponsored byAncestry.com, the genealogical research site.

The first mention in the Deseret News of a battleship to be named for Utah was in May 1903. President Theodore Roosevelt and Navy Secretary William Henry Moody were in Utah's capital, speaking in the Salt Lake Tabernacle.

Following Roosevelt's address, Utah Gov. Heber M. Wells said Moody also would speak, and he mentioned that "some day we may want him to name a battleship Utah." The paper reported, "This sentiment took immediately with the audience, which cheered enthusiastically."

Some six years later, on March 9, 1909, work began on the USS Utah at the Naval Shipyard at Camden, N.J., under the symbolic sponsorship of Mary Alice Spry, daughter of then-Utah Gov. William Spry, according to the Web site honoring the ship and crew,www.ussutah.org. Information in this article is derived from the site as well as from Upton, Hughes and reports and historic photos in the Deseret News archives.

On Dec. 23, 1911, at precisely 10:53 a.m., Mary Alice, then 18 and resplendent in a white fur-trimmed coat, recited the standard words: "I christen thee Utah. Godspeed." And the ship slid into the water to begin its varied career.

Ronald Fox, a North Salt Lake resident, chose the photographs from the paper's archives. A collector of political and Utah history items, Fox is one of the organizers of the USS Utah centennial.

The completed battleship was commissioned at the Philadelphia Navy Yard across the Delaware River from Camden on Aug. 31, 1911. It drew approximately 28 feet of water, weighed 21,825 tons and carried a crew of more than 1,000.

In 1914 it landed a battalion at Vera Cruz, Mexico, in an American show of force. During the fighting that followed, according to the site, seven of the crew earned Medals of Honor. A few years later, the ship helped protect American convoys crossing the North Atlantic during World War I.

The naval disarmament treaty of 1922 required the USS Utah's conversion to a target ship. From then on, it served as the target for bombing and submarine attacks, including dive bombing, torpedo bombing and high-level bombing. Although dummy munitions were used, they were capable of penetrating steel decks, and thick timbers were erected to protect crewmen.

USS Utah also did duty as a machine-gun school. On its last sea voyage, it sailed for Hawaii in August 1941. It was a target ship for bombing exercises until that morning of Dec. 7, when it was among the first ships to be hit by Japanese torpedo bombers in the attack.

Moored at the west side of Pearl Harbor's Ford Island, USS Utah rolled over and sank. It was first hit about 8:01 a.m., and by 8:13 it had capsized. Six officers and 32 enlisted men were lost, while 30 officers and 431 enlisted men survived.

Upton was a Radioman 3rd Class at the time. He was preparing to go ashore to visit Waikiki Beach when the attack began.

"I was just reaching over my locker to get my shaving gear [when] the first torpedo hit us. We didn't know what it was," he said. It was a terrific impact, according to Upton.

The radioman's berth was next to the main radio, two flights below deck, he recalled. The radioman on watch said it was probably a collision by another ship. But within half a minute, "the second torpedo hit us."

The ship started to list to port (left, for landlubbers), and the men realized they needed to get up the ladders so they could go overboard.

With the ladders crowded with desperate crewmen, it took a while to climb up. At one level, someone was lowering a heavy battle grating over the exit, which would have trapped the men below, but he was talked out of it.

"We finally came up to the main deck," Upton said. They found themselves just under the ship's superstructure, with Japanese planes strafing sailors. The ship was listing more, and the protective timbers were bowed.

"I was quite sure we were going over by that time. I slid down the bottom of the ship. There was debris and other men in the water down below, too risky to jump." He was 60 or 70 feet above the water at that point. Mooring lines were snapping as the ship heeled further.

"I was on the bottom of the ship as she rolled over," Upton said. He slid about 40 feet on the hull.

"And sliding down here, there were barnacles around and I was beginning to scratch my legs a bit. When I got down to the bilge keel, kind of a stabilizing keel, I jumped into the water from there."

He plunged 10 or 12 feet into oily water. Luckily, the scum was not on fire at that location, though leaking diesel fuel and oil burned elsewhere in Pearl Harbor.

Upton swam to a mooring quay, helping a lieutenant reach safety. Men from the Utah climbed the quay's cross timbers. Then a ship picked up the survivors and took them to a dock on Ford Island.

"Most of us went into a trench being dug over there for laying a pipe or something. We were in there for a while and saw quite a bit of the action from there."

Asked to describe the sounds, he said, "There were bombs exploding and then there were quiet times, and then they would start in as different planes came back and strafed. It didn't take the old Utah long to capsize. She was on the bottom."

They could see a heroic rescue party headed by Machinist Stanley A. Semanski return to the ship during the strafing. The team cut a hole in the bottom of the ship to rescue Fireman

2nd Class John Vaessen. "He took a torch and cut a hole in it," Upton said. "He (Vaessen) was the only one rescued after the ship had capsized."

Today, he said, Vaessen lives in San Mateo, Calif.

A flatbed truck drove the survivors to a storage building at the Ford Island Naval Air Station, where they unloaded gear such as gas masks, helmets and ammunition. "While we were in there, the second attack came and (the Japanese) started bombing and strafing again.

"Some of the concrete that was blown in the air came right through the roof, and then they started strafing." Machine-gun bullets blasted through holes the concrete had made in the corrugated steel roof. "I think that was the closest they came to getting me," Upton said.

Altogether, America's first casualties of World War II were 2,403 killed, including 68 civilians, and 1,178 wounded, according to Naval Station Pearl Harbor's Internet site. Twenty-one Navy ships sank or were damaged, including all eight battleships in the harbor. Many aircraft and facilities were destroyed at army air fields.

Some of the ships were raised and refitted while others, like the USS Arizona, were too badly damaged. Utah's namesake ship remains next to Ford Island. The naval station site says it was "considered not worth the effort" of repairing.

Upton still keeps in touch with his old shipmates, often by e-mail.

He expressed regret that he could not attend the March 9 reception and display, but he said of the centennial observance, "I think that's great."

By Joseph Bauman
For the Deseret News
Published: Monday, March 2 2009 12:00 a.m. MST

34

Warren at Daughters of the American Revolution Santa Clara, CA

"Pearl Harbor Survivor Shares Tale of Being Aboard USS Utah" Cupertino Patch

By ANNE ERNST (Open Post) November 15, 2011

Warren Upton was on the USS Utah when it was attacked in Pearl Harbor on Dec. 7, 1941. He shared details of that day with the Daughters of the American Revolution.

The morning of Dec. 7, 1941, Navy man Warren Upton reached over to his locker to get his shaving gear out when the ship he was on in Pearl Harbor was hit. Minutes later it sunk.

"I was going to shave and get ready and go swimming over on Waikiki or something, I thought," Upton said.

He was invited to speak to the Santa Clara Chapter of the Daughters of the American Revolution at a luncheon at Blue Pheasant Restaurant on Veterans Day.

The day World War II began Upton was a 22-year-old Radioman, Third Class, aboard the U.S.S. Utah, which was moored off Ford Island at the time of the Pearl Harbor attacks. He had been in the Navy about 22 months when the attack occurred. While the attack on Pearl Harbor lasted a couple of hours it took only minutes for the U.S.S. Utah to list, and 58 men to perish.

"Well about five minutes of eight a terrific explosion shook the ship, and it was within seconds the second explosion (hit,)" he said.

Upton was among 461 survivors, including one that was rescued from the hull of the ship.

The Utah was an "old battleship, it was experimental," Upton said.

"They used us for bombing targets, they used us for anti aircraft for gunnery school for the fleet, and they could put us under complete radio control, remote, everything would work by radio signal. Very few people knew that at the time. It was hush-hush," he said.

Out of 60,000 military personnel stationed in Pearl Harbor at the time, an estimated 2,500 to 3,000 remain, according to Pearl Harbor Survivors Association. The dwindling number of survivors shrinks more, still, when physical health and cognition are factored in.

It's precisely why Robin Hurwitz, webmaster for the DAR chapter, brought her daughters with her to the luncheon.

They may never have a chance to hear someone like this again, she said.

Upton's recall of the day, those short minutes of being under attack and the hours and days that follow, is full of detail. He could see dive bombers through the openings of the doors, he used a gas mask as a pillow and his underwear as a towel after his first shower after the attacks, and the scenes of where wounded were receiving medical attention.

"I can remember going by one room that had casualties and some lady in an evening gown was tending the wounded," he said.

Transcript of 3 minutes of 30 minute talk...

I was getting ready to go ashore on Sunday, the December 7th. No inkling of any sign of enemy, naturally. I was just reaching over to my locker to get shaving gear out and I was going to up to shave and got ready to go swimming around Waikiki or something I thought (laughter). Well, about five minutes of eight a terrific explosion shook the ship and it was within seconds, (a) second explosion, the ship I was on was the USS Utah. It was an old battleship, it was experimental and they used us for bombing target, they used us for anti-aircraft gunnery school for the fleet and they could put us under complete radio control, on remote. Everything would work by radio signal and very few people knew that at the time. It was hush-hush.

    We knew by the time we got up to the main deck that something was up. The bullets from the strafing were ricocheting down the ladders. And I tell you, we were actually hit by two torpedoes from a squadron that came in from the west. By that time, the ship was really listing. There were some life-jackets stowed up. I didn't get any, I could swim. I went right over the side. I don't remember of any lifelines being up and I took of my shoes and my wrist watch and for some reason pitched them into the water (laughter). I slid down across the barnacles and got scratched up. The uniform of the day was white shorts….

Surviving U.S.S. Utah Crewmembers Unite in Laughlin

By LEWIS CLEVENGER Mohave Daily News: Local News
Posted: Saturday, May 19, 2001 12:00 am

(In photo: Surviving crew members of the U.S.S. Utah and their families gathered at the Ramada Express Hotel & Casino in Laughlin recently for the annual reunion of the U.S.S. Utah Survivors' Association. The battleship was torpedoed and sunk, killing 57 of her crew, during the Japanese attack on Pearl Harbor. Pictured above are (from left) Charles Streeter, Abe Lincoln, Lee Soucy, Warren "Red" Upton, William Hughes, John Finn and Guy Pierce. Finn is the only surviving Medal of Honor winner of the 15 awarded for gallantry during the attack. Another member of the crew, John "Jack" Vaessen, was unavailable for the photo. Vaessen, who was trapped inside the ship when it turned over, was the only member trapped inside who was saved by a crew cutting through the hull.)

LAUGHLIN -- Like other Pearl Harbor survivors, the surviving crewmembers of the U.S.S. Utah know they are heroes.

Some were recognized for their bravery shortly after Dec. 7, 1941, while the courage and tenacity of others went unheralded. Some survived the holocaust at Pearl Harbor only to die fighting for their country, while still others survived the war and went on to successful lives afterward.

Yes, they were all heroes, whether sung or unsung, but they were also something else that "Day of Infamy." To a man, they will tell you they were angry, scared, confused and even awestruck at the damage wreaked on the Pacific Fleet by the Japanese torpedo and dive bombers that struck that day.

Recently, several survivors from the Utah gathered at the Ramada Express Hotel & Casino in Laughlin with the families of shipmates since departed to celebrate the annual U.S.S. Utah Survivors. Also attending was John Finn, the sole surviving Medal of Honor winner, out of 15 that were awarded that day, 10 posthumously.

The memories of that day are still sharp in the survivors' minds.

Guy Pierce, who had just turned 17, remembers it well. Holder of the Navy Flying Cross, the Silver Star, and two Purple Hearts, Pierce was barely out of boot camp. Later in the war his plane was shot down over the island of Luzon and he spent 13 months as a prisoner of the Japanese.

But at the start of the war, Pierce was lying in his hammock below deck and contemplating a trip to the ship's mess for a cup of coffee, when he was jarred out of his hammock by an explosion on the left side of the ship. Scant seconds later, a second torpedo struck the same side and the huge battleship began to list seriously.

"As she heeled over, I found myself hanging on to a ladder, wondering what was happening," Pierce recalled. "Then I noticed a group of sailors heading topside for the starboard (right) side of the ship, so I figured I'd better head that way, too."

Dodging rolling timbers covering the ship's deck as protection against dummy bombs from friendly airmen taking advantage of the aged battlewagon's latest role as a target ship, he worked his way across to the other side. As he came out on the main deck, Japanese aircraft dove in to strafe the dying ship, he said.

"I saw a lieutenant firing at the Japanese planes with his .45 (pistol)," Pierce said. "The planes just cut him in half."

Clambering over the protective blister on the starboard side, he began scuttling crabwise over the side of the ship. However, the moss and seaweed clinging to the hull that had been underwater made footing treacherous, and he plunged into the water, eventually making it to shore.

William Hughes and Warren "Red" Upton were in the comfy quarters of the radio operators' bunk room adjacent to the main radio room. Hughes was making preparations to shave, and Upton, who had gotten up earlier to prepare to go ashore, was looking for something in another's foot locker when the first torpedo struck.

"One of the guys said we must have been rammed by another ship," Hughes recalled. "Within 20 seconds, another explosion hit the left side and the Utah began listing to port. It was obvious we needed to reach topside immediately."

Once on deck, the group took temporary shelter from the strafing planes under the ship's superstructure.

"Upon reaching the main deck, it sounded as though all hell had broken loose," Upton said. "By this time, almost everyone was aware of what was happening. We were being attacked by Japanese planes."

Moving over the side, the men then made their way to the beach and took cover in a freshly dug pipeline trench, from where they watched the remainder of the battle.

Of all the stories involving the survivors, however, none is more thrilling than that of John "Jack" Vaessen. Vaessen holds the distinction of being the sole sailor rescued of the 58 trapped below decks after the ship rolled completely over.

Vaessen said after he realized he was trapped, he grabbed a wrench and a flashlight and squirmed into the bilge, which was now the top, where a pocket of air had been trapped. Tapping with the wrench, he suddenly heard an answer from the outside.

"I got an answer, then silence, then rat-a-tat-tat," Vaessen said. "I thought that was a pneumatic tool, (but) it was strafing (from the Japanese planes), and the rescue crew ducked behind the bilge keel."

Outside, two sailors who heard his tapping scurried to a nearby cruiser, the U.S.S. Raleigh, to get help. Though their own ship was badly damaged, a rescue crew set out and, with a torch, cut through the hull and pulled Vaessen to safety.

The Forgotten Ship

Former USS Utah crew members recall lesser known battleship in Pearl Harbor

By Amy O'Donoghue Deseret News

Published: Sunday, December 6, 2009 12:00 a.m. MST

One hundred years ago she was the belle of Utah's ball, the largest and most powerful battleship in the world, launched in New York on Dec. 23, 1909 with these words: "I christen thee Utah! Godspeed!"

Thirty-two years later on Dec. 7, two torpedoes struck the USS Utah at Pearl Harbor and within minutes she rolled over on her side, taking some of her crew to rest with her forever.

As the decades passed, a solemn tribute to the horrific nature of the attack and the sacrifices of that infamous day was erected over the tomb of the USS Arizona, which would become the Pearl Harbor National Monument.

But on the opposite side of Ford Island rests the rusted hull of the USS Utah, which over the years has been termed "The Forgotten Ship."

Four men who have lived a total of 349 years on Earth ventured to the state of Utah earlier this year to remember what many have forgotten.

In the halls of the state Capitol, where an exhibit celebrates the life and times of the USS Utah, they talked of how good it was to see each other again, and how few they have become.

The four are among 30 or so men left from the 461 survivors of the assault on the 21,825-ton ship.

They felt her shudder from the torpedoes and watched tables, chairs, salt shakers and sugar bowls slide across the floor when she began to list.

"And it was one hell of a day, a terrible day," said Cecil Calavan, now 85.

Calavan was a 17-year-old farm boy in 1941 practically living on his own because of a family split. He talked his father into letting him sign up for the Navy, which had started recruiting 17-year-olds a year earlier.

He had grand visions of serving aboard a glorious battleship.

"When I first saw the Utah, I almost cried," he said. "She'd been cut back, she didn't look like a battleship anymore."

By the time Calavan came along, those glory days as a battleship were behind the USS Utah, which had seen action off the coast of Mexico in 1914 and was assigned to Bantry Bay, Ireland, to protect British interests at the beginning of World War I.

When Calavan came along, it had been a few years since the ship was selected to take President-elect Herbert Hoover and his wife on an extended tour of South America.

In 1941 when Calavan stepped aboard, the USS Utah had already spent the last decade as training and target ship for new bombers, barely reminiscent of her fighting past.

After only a couple of months on the Utah, Calavan changed his mind about her.

"I knew before she went down she was a great ship."

At the time of Pearl Harbor, Calavan was a seaman 2nd class, making $36 a month.

He was shaving with his Schick Injector — the newest thing on the market in 1941 — when he heard a plane come screaming in.

"It was a tremendous roar. I had heard planes overhead before, but nothing screaming like this. Within two minutes, the ship shuddered."

He looked out the porthole, and the next torpedo hit.

"I still could not believe it was an attack. We could see what we called the 'flaming hemorrhoids' on the planes and I knew."

By then, the USS Utah was in chaos. Seaman 2nd Class Robert Ruffato, now 86, said the men in the engine room were scrambling up to get out, and the men topside were trying to get below to escape the strafing by Japanese aircraft.

As the Utah began to list, Ruffato and a buddy rushed to the forward part of the ship, and dodging sliding furniture, jumped into the harbor.

The water was being riddled with machine gun fire.

"I'd go about 10 or 12 feet down, hang onto the corral and you could hear the bullets hitting the water, they were hot and sizzling."

Ruffato held his breath as long as he could, swam up, took a deep breath and went back down.

It was only 50 yards from the shore, but Ruffato said three 18-year-olds who didn't know how to swim were killed as they struggled in the water.

Scratched up and bleeding after sliding down the barnacles on the Utah, Warren "Red" Upton found himself bobbing in the water next to his lieutenant, who was clad only in his pajamas. "He hollered at me, 'Can you swim, Red?'"

Upton, now 90, helped the officer over to the pilings and a motor launch ferried them to the dock, where the men took cover in a ditch.

"We were there for quite a few minutes, got to see quite a bit of the action," he said. "It was like a kaleidoscope of events is the way it impressed me, changing from one thing to another so rapidly. Naturally, you're trying to save your skin."

There in the ditch, he was just beginning to absorb the violent shaking of the Utah, of ducking ricochetting bullets as he scrambled up the starboard ladders. That night, he'd bunk with a bunch of other stranded sailors on the USS Argonne, using a gas mask as a pillow.

It was only a few days later, bandaged and lying in a hospital bed, that the enormity of what happened settled in for Clark Simmons, now 88.

A 20-year-old officer's steward from Brooklyn, Simmons had been on the Utah for a little over two years at the time of the Pearl Harbor attack.

"The Japanese were kicking our butts. It was something that we were not trained to do. We weren't trained for attack. We weren't at war."

Hit by a piece of shrapnel, Simmons impatiently convalesced in a hospital bed, listening to a replay of President Franklin D. Roosevelt's speech to the nation.

"It was at that time I realized how significant and serious the attack was. Some of the fellows visited the hospital, told me exactly what happened, the people who had been killed. It was really frightening."

It seemed like a lifetime ago that Simmons had ventured to shore — or the beach as the men called it — on the night preceding the attack to shop for papaya and pineapple. In the days to follow he and the others would find homes on new ships.

"If you lived through that attack, everybody had to fight now. The war was on," Calavan said. "Over the years, even nice men will fight."

Accounts differ, but the most widely agreed on number is 58 — 58 men who never made it off the Utah and remain with her today. The USS Arizona, which erupted in a fiery explosion, took 1,177 of her crew and remains a permanent memorial to the attack. The USS Oklahoma lost 429 sailors and Marines, but two years later she was righted and the bodies were recovered for burial. While being towed to California to be used as scrap, the Oklahoma sank at sea, leaving the Arizona and the Utah as the only ships at Pearl Harbor to remain at their berth.

It took 31 years, but a memorial was finally dedicated for the USS Utah, about a mile from the Arizona.

The surviving Utah crew figure that the horrific number of lives lost on the Arizona, coupled with limited public access to the Utah, has made it all too easy over the years for the dreadnought to earn the moniker of "The Forgotten Ship."

In 2008, however, one of the last things President George W. Bush did was sign an executive order transferring jurisdiction of the Utah from the U.S. Navy to the National Park Service. The USS Utah Association anticipates that someday; the park service will provide transportation to visitors via shuttle to see the Utah.

There are some survivors, however, who reject the notion of ever having it duplicate the style of the Arizona memorial. As one Utah survivor once told a friend, the Arizona is like a gorgeous, solemn church, and rightfully so.

"But everyone should have to see the Utah. It shows what war is actually like. The other shows what man can build."

By Amy O'Donoghue Deseret News

e-mail: amyjoi@desnews.com

Photos

Figure 1: USS Utah at Pudget Sound 19AUG1941

Figure 2: USS Utah under steam.

Figure 3: Deck of USS Utah with large timbers to protect from training bombs with paint.

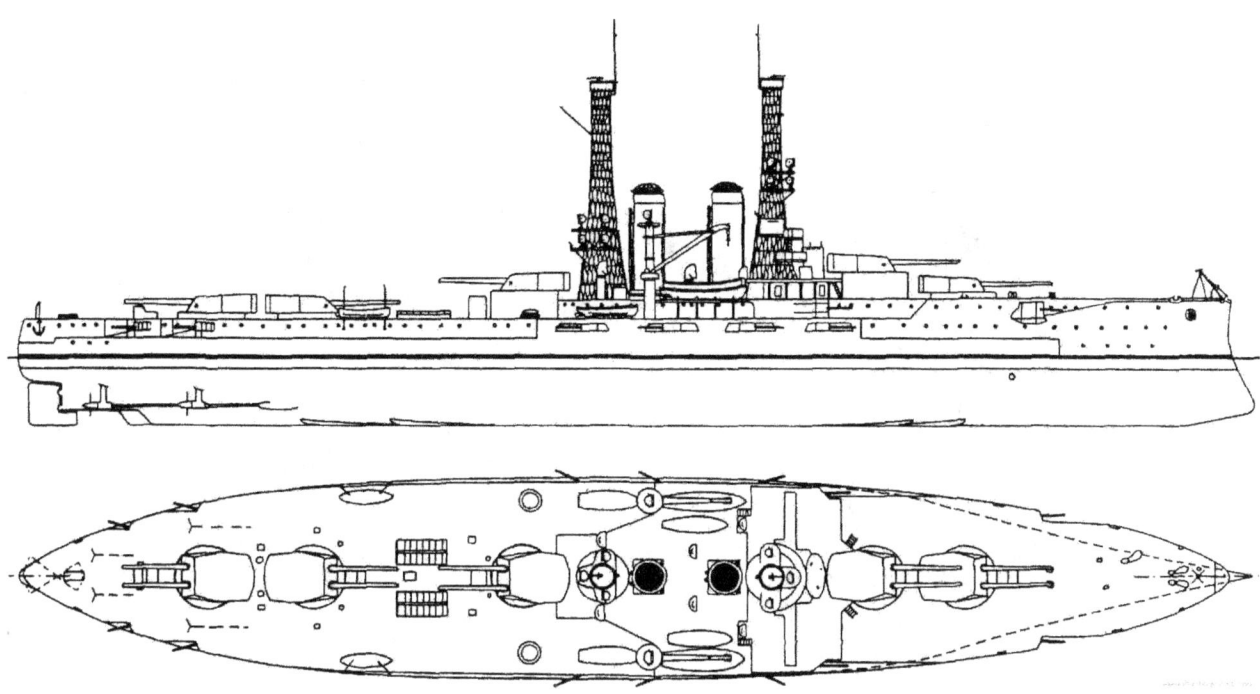

Figure 4: Blueprint of the USS Utah.

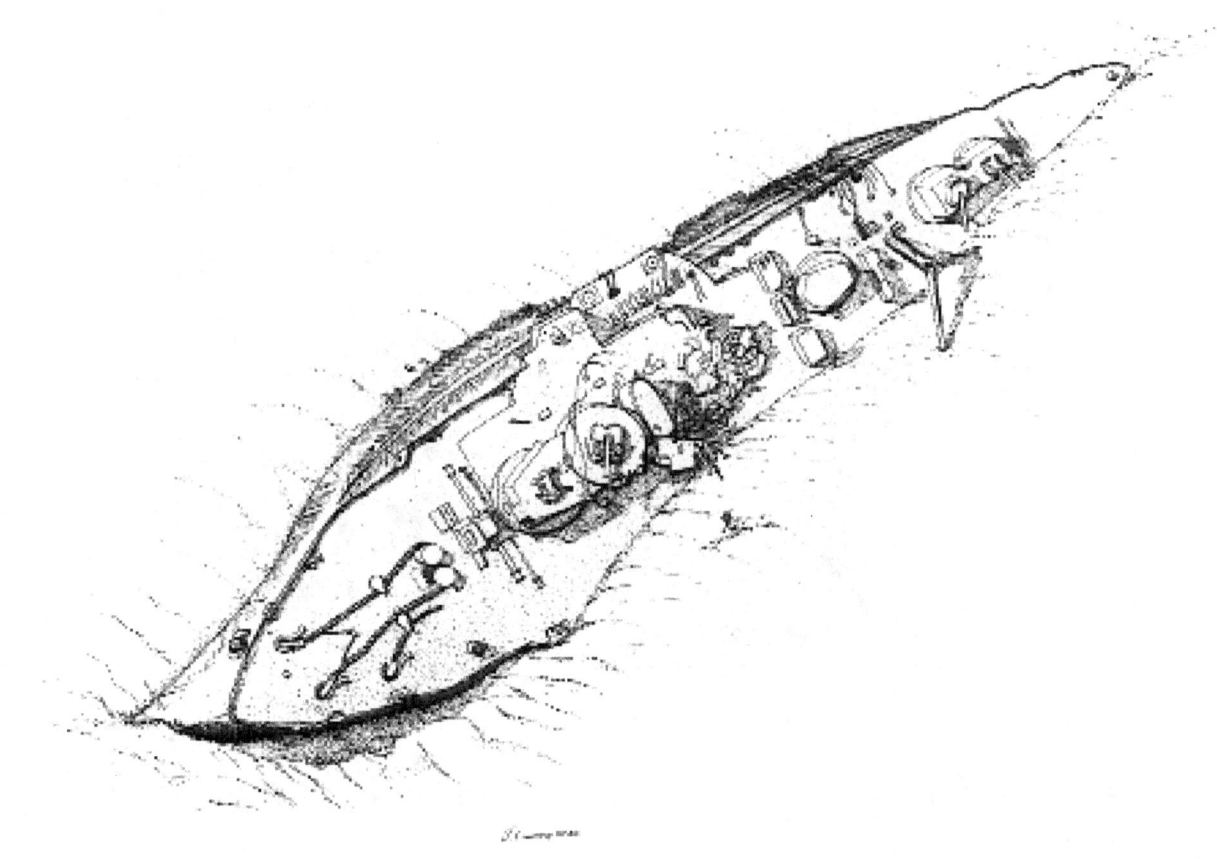

Figure 5: Sketch of final resting position of USS Utah.

Figure 6: Composite montage of before & after aerial photographs showing carriers in port.

Figure 7: Aerial photograph of overturned USS Utah.

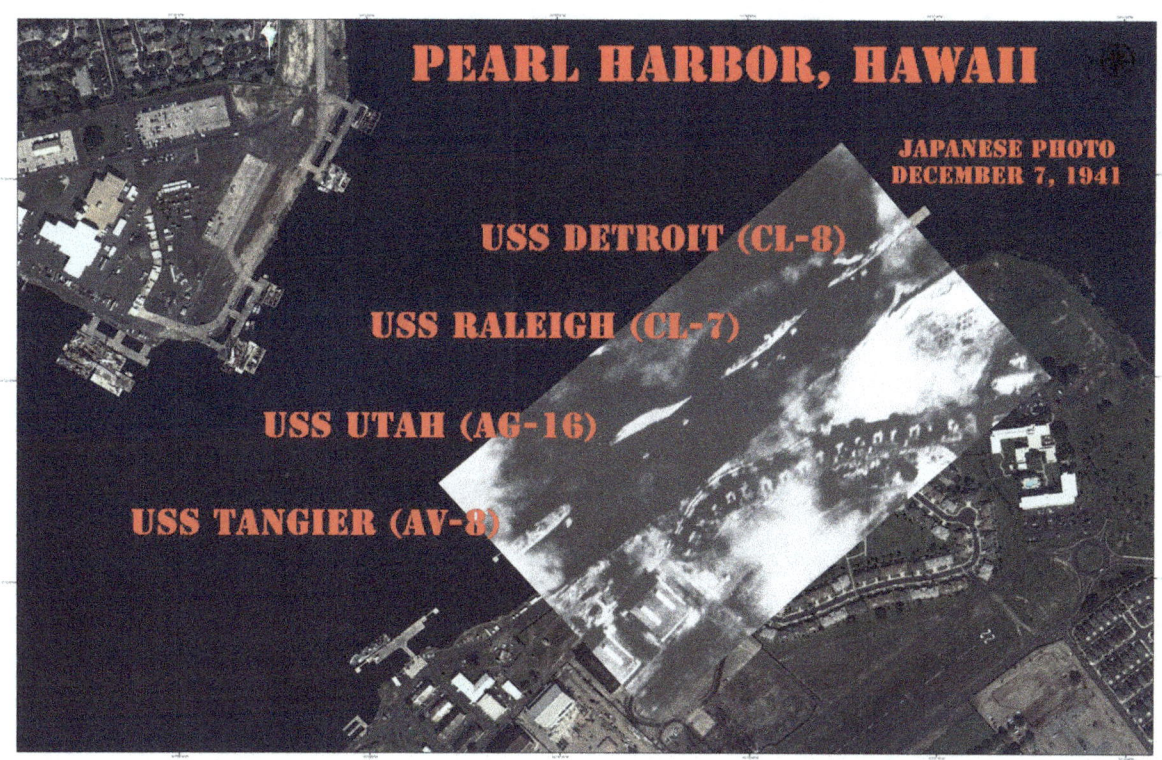

Figure 8: Map of NW side of Ford Island with overlay of 1941 aerial photo.

Figure 9: Map of NW side of Ford Island.

Figure 10: USS Utah survivor Warren Upton, 100, left, was greeted Friday by Burke Waldron, 96, from the state of Utah, who served with the U.S. Naval Ground Force Pacific. Struck by torpedoes, the USS Utah capsized and sank within 12 minutes. BRUCE ASATO / BASATO@

Figure 11: Warren Upton, 92, of San Jose, was aboard the USS Utah, 70 years ago, and was at Pearl Harbor Day, at Moffett Museum in Mountain View, on Wednesday, Dec. 7, 2011. (Karen T. Borchers/Mercury News).

Figure 12: Pearl Harbor Survivor and USS UtahRadioman Warren Upton and wife Jean with Eagle Scout Christian De Knikker in Woodland, CA, April 18, 2009. Photo. Woodland Record [woodlandrecord.com] Article Monday, May 04, 2009.

Figure 13: Warren at a meeting of the Pearl Harbor Survivors in San Jose, CA.

Figure 14: Warren and great-grandsons.

Figure 15: Warren and his Grandson.

Figure 16: Warren and his daughters.

Figure 17: Warren and his family.

Figure 18: Blowing out the candles for 100th birthday.

Figure 19: Pearl harbor Survivor Mickey Ganitch (USS Pennsylvania)

Index

barnacles, 7, 8, 22, 30, 34, 40
battles, 3, 6, 7, 11, 22, 30, 36
battles & invasions
   Eniwetok, 11
   Korea, 12, 15
   Makin, 10
   Pearl Harbor, i, iii, 2, 3, 4, 5, 10, 20, 21, 28, 29, 30, 31, 33, 35, 39, 40, 41, 48, 49
   Philippines, 15
biking, 6
birthdays, 2, 52
Boy Scouts, 49
bruises, 8
business
   American Institute of Banking, 17
   banking, 17, 18
   Beckman Instruments, 18
   Bureau of Automatic Repair, 23
   Crocker Bank, 18
   housewife, 3
   investigator, 23
   investments, 18
   jobs, 3, 11, 12, 15, 16, 19, 23
   Mackay Radio, 12, 17
   nursing, 24, 26, 27
   philosophy, 23
   phone company, 10
   police academy, 19
   schooling, 2, 3, 4, 5, 17, 24, 26, 27, 30, 33, 34
   stock brokers, 18
   YMCA, 6
Chief Petty Officer, 14
coffee, 9, 35
collision, 6, 30
crash-land, 9
cruises, 12
dates
   1909, 29, 39
   1911, 29
   1922, 30
   1932, 5
   1939, 3
   1940, 4
   1941, 29, 30, 33, 35, 39, 40, 47
   1946, 17
   1958, 19
   1988, 20, 29
   2015, vii, 3
   2019, 3
   Christmas, 3, 5, 10, 11, 13
   Christmas Eve, 10
   December 7, 1941, 6, 20, 33
   December 8, 1941, 9
economics, 17
family, 2, 3, 12, 17, 19, 20, 23, 25, 26, 27, 29, 50
   brother, 3
   children, 3, 20, 26
   Chuck, 23
   daughters, 33, 51
      Diane, 1
      George, 41
   grandchildren, 26
   granddaughter, 28
   grandkids, 19
   Haley (granddaughter), 28
   Jean (wife), v, vii, 3, 12, 16, 24, 49
   Mary (daughter), 23, 29
   mother, 3, 26
   Roxie (dog), 27
   sisters, 1, 3, 10, 17
   uncle, 3
   Warren, i, iii, iv, vii, 3, 15, 16, 21, 22, 23, 25, 29, 33, 35, 36, 40, 48, 49, 50, 51
grease, 8, 14
leaders
   President Franklin Delano Roosevelt, 16, 29, 41
marriage, 19, 24, 25
military
   Army, 5, 25
   bands (Navy ships), 5
   barracks, 8
   base
      Alameda, 3, 17

Alameda Naval Air Station, 3, 17
bases
    Clark Field, Philippines, 40
    NAS Moffett Field, 48
    Naval Air Base San Pedro, 25
    Naval Air Station Cecil Field, Florida, 1, 39
    Naval Base Pearl Harbor, i, iii, 2, 3, 4, 5, 10, 15, 20, 21, 28, 29, 30, 31, 33, 35, 39, 40, 41, 48, 49, 52
    United States Fleet Activities Yokosuka, Japan, 16
bombardments, 11
bombing, 5, 22, 30, 31, 33, 34, 35, 36, 39, 44
classification tests, 4
commissary, 7, 8
commission, 15
convoy, 10
duties
    cadet officer, 3
    enlisted, 4, 12, 30
    enlistment, 12, 17
    guard, 14
Echo Taps, 21
equipment
    4th Marine Defense Battalion, 10
    armor-piercing shell (armour-piercing), 9
    atomic bomb, 16, 22
    Browning, 4
    drones, 13
    dungarees, 8
    five-inch gun, 9, 22
    Geedanus (sp?) stand, 7
    gun, 4, 5, 9, 30, 31, 40
    gunneries, 5
    gunnery, 5, 10, 33, 34
    guns, 4, 9, 10, 21
    headphones, 13
    helicopters, 14
    jets, 9, 14
    Lewis automatic machine gun, 4
    rifles, 4
    torpedoes, 5, 14, 29, 30, 34, 35, 36, 39, 40, 48
    Underwood typewriters, 13
    uniform, 7, 34
    weapons, 16
    World War I plane, 13
Golden Dragon Award, 11, 14
Japanese, 7, 16, 23, 29, 30, 31, 35, 36, 40
Japanese plane, 29, 30, 36
joining Navy, 3, 12, 24
Korean War, 15, 16, 17, 18
Marines, 10, 41
Navy, 3, 4, 5, 9, 10, 12, 14, 16, 17, 18, 20, 21, 24, 25, 26, 27, 28, 29, 31, 33, 35, 39, 41, 48
organizations
    4th Marine Defense Battalion, 10
    Bureau of Economic Repair (BAR), 23
    merchant marine, 3
    Naval Academy, 4, 7
    Navy District 14, 9
Pearl Harbor ashes, 21
Pearl Harbor bicycle, 6
personnel
    Admiral Chester Nimitz, 15
    Admiral Halsey, 15
    boot camp, 4, 35
    Chief Galli, 9
    Chief Petty Officer, 14
    Chief Putman (sp?) Chief Radioman, 9
    chief radioman, 8, 9
    chief tour captain, 4
    Chief Warrant Officer, 16
    chiefs, 4, 8, 9, 10, 16, 17
    Commandant, 9
    Commander, 8, 21
    Commander of the Hawaiian Navy Region, 21
    Ensign, 15
    general, 11, 13, 16
    General Douglas MacArthur, 15
    generals, 15
    infantry, 4
    midwatch, 9
    President Eisenhower, 15
    President Franklin D. Roosevelt, 16
    President Harry S. Truman, 17
    Ray Hanson (friend), 6
    recruit, 4
    recruiters, 3
    Senior Petty Officer, 10

shipmates, 2, 12, 17, 19, 20, 29, 31, 35
showers, 8, 33
signalman, 5
striker, 9
train trip (Yokosuka-Kyushu), 16
watches, 4, 5, 6, 8, 9, 10, 11, 12, 13, 23, 30, 34, 36, 39
watchmen, 5
yeoman, 9
radio
    blinker-light, 5
    call sign NAFT (California), 13
    call sign NEDJ (West Virginia), 13
    call sign NIQJ (Utah), 13
    code, 5, 13
    electronic, 13, 15
    electronics, 3, 13, 18
    frequency, 12, 13
    international distress frequency, 10
    Morse code, 13
    radio, 5, 9, 12, 13, 15
    radio shack, 6, 8, 10
    semaphore, 5
    ships bridge, 9, 13, 14
    telegraph, 13
    teletype, 17
    transmit, 13
    transmitter, 13
    vacuum tube, 13, 16
reunion
    60th reunion, 21
    Mount Diablo Beacon Lighting, 16
    reunions, 2, 20, 21, 25, 26, 29, 35
    USS Utah Association, 19, 29, 41
    Veterans of Foreign Wars (VFW), 20
ship bunks, 6
strafing, 7, 29, 30, 31, 34, 36, 40
submarine, 10, 30
military personnel
    Japanese, 7, 16, 23, 29, 30, 31, 35, 36, 40
    lieutenant, 7, 9, 30, 36, 40
    Marines, 4, 5, 10, 25, 41
    nurses, 16, 24, 25, 27
    officer, 5, 7, 8, 12, 17, 23, 40
    officers, 5, 9, 12, 30
    pilots, 5, 7, 9, 14
    prisoner, 35
    radioman, 6, 11, 12, 14, 30
    seaman, 9, 40
    squadrons, 5, 34
oil, 8, 30
Pearl Harbor Survivors, i, iii, 33, 35, 49
people
    17-year-old Chuck, 39
    Amelia Earhart, 14
    Caldi Nicker, 20
    Gross (striker), 9
    President Harry S. Truman, 16
    teachers, 3, 28
    Ted Ivy, 20
pinball machines, 17
places
    Africa, 24
    cities
        22 Battery Street in San Francisco, 17
        Berkeley, California, 9, 15, 17
        Cresent City, California, 10
        Dallas, Texas, 29
        El Dorado, California, 3
        Forest Grove, Oregon (birthplace), 24
        Fremont, California, 3
        Hollister, California, 28
        Honolulu, Hawaii, 2
        Incheon, South Korea, 15
        Kelso, Washington, 26
        Kyushu, Japan, 16
        Las Vegas, Nevada, 19
        Laughlin, Nevada, vii, 20, 35
        Mare Island, Vallejo, California, 10
        Montgomery Street, San Francisco, 18
        New York, New York, 39
        Oakland, California, 17, 25
        Palo Alto, California, 18
        Portland, Oregon, 24, 26
        Redwood City, California, 4, 17
        San Diego, California, 4
        San Francisco Bay Area, California, 3, 4, 9, 10, 12, 16, 17
        San Mateo, California, 18, 19, 31
        Santa Clara, California, vii, 3, 33
        Santa Cruz, California, 12
        Shoemaker, California, 16
        Warm Springs, California, 3
    countries

Japan, 16
Kingdom of Tonga, 10
Korea, 12, 15
Nigeria, 24
North Korea, 14
Philippines, 15
South Korea, 12
United States, 1, 3, 9, 10
country states
    Luzon, Philippines, 35
Diamond Head, Oahu, Hawaii, 6
Europe, 25
Golden Gate Bridge, San Francisco, 10
Hoover Dam, Arizona-Nevada, 16
International Date Line, 11, 14
islands
    Bora Bora, French Polynesia, 10
    Eniwetok, Gilbert Islands, 11
    Espiritu Santo, Vanuatu, 11
    Ford Island, Oahu, Hawaii, 7, 9, 30, 31, 33, 39, 47
    Gilbert Islands, 11
    Makin Island,Kiribati.(Gilbert Islands), 10
    New Hebrides Condominium, 11
    Pearl Harbor, Oahu, Hawaii, i, iii, 2, 3, 4, 5, 10, 20, 21, 28, 29, 30, 31, 33, 35, 39, 40, 41, 48, 49
    Republic of the Marshall Islands, 11
    Wake Island, Micronesia, 9
Japan, 16
Middle East, 16, 23
Oakland, 17, 25
Oakland Naval Hospital, Oakland, California, 25
Okinawa, Japan, 12
Pacific Ocean, 1, 11, 16, 35, 48
Pearl Harbor, i, iii, 2, 3, 4, 5, 10, 20, 21, 28, 29, 30, 31, 33, 35, 39, 40, 41, 48, 49
reserve bases:, 29
Sierra foothills, 3
states
    Arizona, 8, 31, 39, 41
    California, 3, 9, 13, 16, 19, 25, 41
    Hawaii, 1, 5, 6, 21, 30
    Maryland, 13
    Oklahoma, 2, 41
    Oregon, 24, 26, 27
    Texas, 20
    Virginia, 13
The Orient, 15
Treasure Island, California, 12
trench (Ford Island), 7, 30, 36
UC Berkeley, Berkeley, California, 15, 17
Waikiki Theater, Honolulu, Hawaii, 6
Washington High School, 3
Washington, DC, 3, 26
West Valley College,Saratoga, California, 23
Yosemite National Park, California, 26
planes
    aircraft, 5, 13, 31, 33, 34, 36, 40
    airlines, 13
    airport, 17
    Douglas Aircraft Company, 29
    Mitsubishi Ki-21 (Sally), 1
religion, 3, 23
school
    elementary, 3
    graduating, 3, 5, 25
seasickness, 5
shaving, 6, 7, 30, 33, 36, 40
ships
    battleships, 5, 29, 31, 33, 34, 35, 36, 39
    battlewagons, 5
    carriers, 14, 46
    destroyers, 5, 9, 10
    midget subs (Japanese), 7, 9
    ships, vii, 1, 2, 5, 6, 7, 8, 9, 10, 13, 14, 15, 16, 21, 23, 25, 29, 30, 31, 33, 34, 35, 36, 39, 40
    submarines, 5, 7
    USS Argonne (AS-10), 8, 9, 40
    USS Arizona, 8, 31, 39, 41
    USS Brazos (AO-4), 4
    USS Burrows (DE-105), 15
    USS Castor (AKS-1), 9, 10, 14
    USS Dixie (AD-14), 11
    USS Enterprise (CV-6), 9
    USS Henley (DD-762), 10
    USS Monaghan (DD-354), 7
    USS Pennsylvania, 52
    USS Saint Paul (CA-73), 15
    USS Utah (AG-16), vii, 1, 2, 4, 5, 7, 8, 9, 12, 13, 20, 21, 22, 25, 26, 29, 30,

31, 33, 35, 36, 39, 40, 41, 43, 44, 45, 46, 48
swimming, 7

telephones, 9, 13
Veterans of Foreign Wars, 20
volunteering, 2

www.ingramcontent.com/pod-product-compliance
Lightning Source LLC
Chambersburg PA
CBHW081348040426
42450CB00015B/3357